D0519284

THE LITTLE BEAN BOOK

Judy Ridgway

PIATKUS

Other titles in the series

The Little Green Avocado Book
The Little Garlic Book
The Little Pepper Book•
The Little Lemon Book
The Little Apple Book
The Little Strawberry Book
The Little Mushroom Book
The Little Nut Book

© 1983 Judy Piatkus (Publishers) Limited

First published in 1983 by Judy Piatkus
(Publishers) Limited of Loughton, Essex

British Library Cataloguing in Publication Data

Ridgway, Judy
 The little bean book.
 1. Cookery (Beans)
 I. Title
 641.6'5'65   TX803.B/
ISBN 0–86188–410–8

Drawings by Linda Broad

Designed by Ken Leeder

Cover photograph by John Lee

Typeset by V & M Graphics Ltd, Aylesbury, Bucks
Printed and bound by The Pitman Press, Bath

CONTENTS

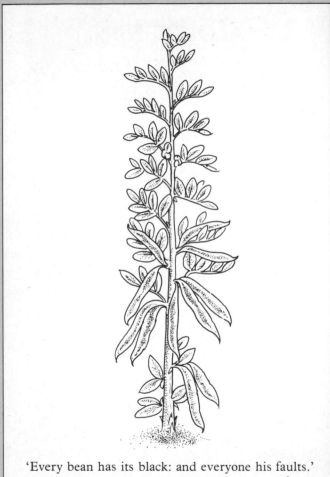

'Every bean has its black: and everyone his faults.'
Anon

THE BEAN FAMILY

The bean family embraces many hundreds of varieties. Strictly speaking, the word 'bean' refers to the pod or seed of certain leguminous plants of the group Fabaceae. They may be eaten fresh, with or without the pod, or shelled and dried. Dried beans, with peas and lentils, are known collectively as pulses.

Most varieties grow either as an erect bush or as a climbing plant, but there are some important beans of intermediate form such as dwarf beans and semi-climbers. Beans grown for their pods tend to produce a relatively low yield of mature beans, often of low eating quality. Those grown for their seeds, on the other hand, often have pods that are too fibrous to be eaten at any stage of their development.

Broad beans are one of the exceptions to the general rule, for they may be eaten very young either in their pods or as freshly-shelled mature beans. These beans are one of the oldest known varieties, originally called Windsor beans. They are very popular in Europe, but almost unknown in America where they are fed only to cattle. Instead, a close cousin, the Lima bean, is grown in the states for human consumption.

Bean pods vary tremendously from variety to variety. They may come in shades of green, yellow, red or purple. They may even be slashed with red and purple. Their shape ranges from flat to round, smooth to irregular and straight to sharply curved, and their length may vary from 3–8 inches or more!

Seeds, too, are just as variable. Solid colours include all shades from white through to yellow and green as well as all shades of tan, brown, pink, red and purple through to black. In addition, there are countless patterns. Shapes range from nearly spherical to flattened, elongated and kidney-shaped.

The soya bean is by far the most important bean in the world. It is used extensively in Eastern cooking, and in the West it is grown both for its oil and protein content. The latter goes into textured vegetable protein products (TVP); soya flour and animal fodder. Soya also goes into products as diverse as paints, chemicals, waterproofing preparations, fire fighting foam and adhesives. Kidney and haricot beans come next on the scale of economic importance with garbanzo beans (chick peas) and broad beans running a poor third. Mung beans are important in the East, where they are used mainly for sprouting.

BEAN NAMES

Some of the names given to beans are extremely picturesque. Oriental beans include the Sword or Sabre bean, the Velvet bean and the Cherry or Cow bean. In the United States, Jack beans are used for green manure, and Lima beans take their name from the fact that they were imported from Lima in Peru.

The Mediterranean Locust or Carob bean owes the first of its names to a faulty translation of the bible. The fruit of the Carob tree is said to have been the 'husks' referred to by Luke, and the 'locusts' eaten by John the Baptist. The latter reference also gives rise to another name for the same bean – John's Bread.

In Britain broad beans used also to be called Horse beans, and this name stuck with the bean when it was taken to the New World with the Pilgrim Fathers. In even earlier times it was known as the Field bean and the Long pod. In Elizabethan England, broad beans were used as fodder for horses, and one of Puck's speeches from *A Midsummer Night's Dream* recalls this fact:

> 'I am that merry wanderer of the night.
> I jest to Oberon and make him smile
> When I a fat and bean-fed horse beguile
> Neighing in the likeness of a filly foal'

A very attractive tree in the bean family is better known as Laburnum, but it is also called Golden Chain or Bean Tree.

Other beans which are not really beans at all include the Tonka bean, which is the fruit of a South American tree whose seeds give rise to oil which is used in the manufacturer of bitters; the Egyptian, Hyacinth or Sacred bean; and the Bean Caper, whose flower buds are eaten as capers.

JUMPING BEAN

Mexico is the home of the most unusual bean of all. Known as the Mexican Jumping Bean, it is the seed of a shrub which contains a Bean Moth caterpillar. These caterpillars eat their way into the seed while it is growing. When the seed falls to the ground the movements of the caterpillar cause the bean to jump about until it gets out of the sun and into the cooler shade. Here, the caterpillar changes into a chrysalis and then into the grown moth. The moth breaks its way out of the (now empty) bean shell and flies away to lay its eggs on another jumping bean flower.

HISTORY OF BEANS

B eans have been cultivated by man since pre-
historic times. The broad bean is the oldest known
bean, but it was not the first to reach Britain. This
honour fell to the Celtic bean which was very much
smaller. These beans arrived in South West England
during the pre-Roman Iron Age.

The Romans had cultivated beans for a long time
before they travelled North. They grew beans for
food, for fodder and to plough back into the soil.
They also used beans as a means of voting in the
elections for their magistrates; the black bean
signified condemnation and the white bean absolu-
tion. The Greeks, too, used the same system.

In Britain crops remained much the same from the
time the Romans left to the Middle Ages, and broad
beans were a staple item. As the manors developed
new systems of crop rotation, broad beans took their
place as field crops, alternating with cereal crops –
hence, perhaps, the name Field beans. Beans were
the food of the common people, and Chaucer
referred to things as being 'not worth a bean'.

In those days bacon was the only meat which the
working man was likely to eat in winter and it was
invariably served with dried beans. The two foods

remained a popular combination for centuries. G. K. Chesterton immortalized this in his poem 'The Englishman':

> But since he stood for England
> And knew what England means,
> Unless you give him bacon
> You must not give him beans.

The great advantage of beans as a crop was that they could be dried and used throughout the year. Writing in the late medieval period, Alexander Neckam refers to 'frizzled beans' for use in stews. These were a French invention, popular in Paris where Neckam had lived. The beans were boiled in their pods, shelled and 'frizzled' in a heated spoon.

The English equivalent was 'canebyns'. The beans were steeped for 48 hours in several changes of water and then hardened in the oven and hulled with a handmill. After this they were broken into pieces and fried. Both the French and the English methods of treating beans were to prevent them from sprouting if they got damp in storage.

As well as being the major ingredient of winter soups and stews, beans also went into medieval bread. Horse-bread, with beans and peas as the official ingredients, was made in the manors and monasteries and even in some of the town bakeries. It is not clear if the name referred to the use of Horse beans or whether the bread was made for the stables. However, in bad harvest years there is no doubt that this bread was eaten by the poorer peasants.

Slightly better quality breads were made by mixing ground beans with cereals such as rye and oats and maybe even a little wheat. But bean flour, even in small quantities, made a nasty tasting bread which did not hold together well.

Bean-meal continued to be used. During the reign of Elizabeth I there were years when bakers were ordered to make their bread of rye, barley, peas and beans, and the townsfolk suffered with the peasants. This type of bread became less common in the 18th century, but even then bean-meal was used by unscrupulous bakers to eke out expensive foreign wheat. This was still happening in London in the 19th century, and indeed it was not until 1872, when the addition of food adulterants was finally banned by law, that poor people could be sure that there were no beans in the bread.

Fresh beans have an equally long culinary history and broad beans were listed in Gerard's *Herball* in 1597. So were French beans, which were introduced into Britain by Flemish and French immigrants in Elizabethan times. Gerard referred to them as Brazil Kidney Beans, and though their origin is not entirely certain they are thought to be native to South America, but to Peru rather than Brazil. They were brought from the New World to France, where they soon became established, thus giving rise to their modern name in Britain. In the 17th and 18th centuries they were even more of a luxury item for the wealthy than they are today.

Runner beans arrived from South America, to be

grown as ornamental climbing plants. It was not until the last century that they were grown for their food value.

As with dried beans, fresh broad beans were daily fare for most people, and even the aristocracy were not above sampling them. The story has it that

George III, when he went to inspect building progress at the Woolwich Arsenal, stopped to eat an *al fresco* breakfast with the workmen. They were having boiled broad beans with bacon, and the king liked the dish so much that he instituted it as an annual breakfast.

The same dish was also a favourite in the great houses of England, where in early summer a side-dish known as Beans and Collops often featured. Bacon would be cooked in wine, spiced with sugar and cloves, and the fresh beans served with parsley sauce. That was the heyday of the broad bean. Since the 18th century it has declined in popularity and French and runner beans have taken its place.

BEAN FACTS AND FIGURES

WORLD PRODUCTION

★ More than 115 million metric tonnes of beans are grown in 88 countries every year – and this figure does not include the abundant harvests of garden-grown vegetables.

★ Soya bean production accounts for more than three quarters of the total figure and, rather surprisingly, the lion's share is grown not in the East but in the United States. Brazil is the second largest soya bean producer, with China in third place.

★ French, haricot and kidney beans account for some 14,000 metric tonnes, and the production tables here are topped by India and Brazil, with the US, Mexico, Africa and China as runners up.

★ India is the main source of garbanzo beans (or chick peas), and China of broad beans. Africa comes quite high in the broad bean production figures, and Italy leads the way in Europe.

★ Europe produces almost half the total figure of 245,000 metric tonnes of fresh green beans. Asia and Africa are the next largest producers, with America a poor fourth. In Europe, Italy and Spain are the top producers, with Britain in third place.

UK PRODUCTION

★ Nearly all the fresh green beans eaten in Britain are home-grown, and only a very small percentage is imported during the winter months. Gross production of French and runner beans stands at 78,000 metric tonnes; broad beans at 57,000 metric tonnes.

★ Broad beans are cultivated mainly in Eastern England, growing as far north as Yorkshire. French beans are concentrated in Southern counties and in East Anglia, and runner beans come from the Vale of Evesham, the home counties and East Anglia.

UK IMPORTS

★ Green bean imports are primarily from Spain, Egypt, Cyprus and Kenya. Indeed, French beans from Kenya are considered by some gourmets to be the very best in the world.

★ Dried beans are nearly all imported and more than 95,000 metric tonnes come in from the United States and elsewhere every year.

UK CONSUMPTION

Averaged out over the year, we only eat 1.88 oz of green beans per person per week, and about a third of these are frozen. Dried beans are consumed at an even lower level.

THE FOOD VALUE OF BEANS

Bean pods, like most vegetables, are low in protein and carbohydrates and contain virtually no fat. They are low in calories, high in dietary fibre and supply Vitamin A, Vitamin C and various mineral salts. French beans contain iron – often defficient in our diet. The Vitamin C content starts to decline as soon as the pods are picked, so they should be eaten as soon as possible after picking or purchase.

Broad beans and dried beans are the mature seeds of the principal beans and are rather similar in composition to each other. They are rich in protein and contain more carbohydrates than other green or root vegetables. They thus provide more energy. Dried beans do not regain all their water when soaked and cooked and so contain twice as much energy, weight for weight, as fresh beans. Beans also provide some of the Vitamin B complex, along with iron, phosphorous and potassium. Fresh beans such as broad beans and lima beans also contain Vitamin C.

Canned beans lose much of their Vitamin B content, but may retain some Vitamin C.

Baked beans are processed in such a way that they lose most of their vitamins, but Vitamins A and C are added with the tomato sauce.

BEAN CALORIE CHART

Calories per 4 oz-portion, cooked

French beans	7
Runner beans	7
Broad beans	43
Canned baked beans	93
Butter beans	93
Haricot beans	87
Kidney beans	90
Soya beans	114
Mung bean sprouts (uncooked)	30
Soya bean sprouts (uncooked)	52

There was an old person of Dean,
Who dined on one pea and one bean;
For he said, 'More than that,
Would make me too fat,'
That cautious old person of Dean.

One hundred Nonsense Pictures and Rhymes
Edward Lear (1869–1944)

BEAN PROTEIN

Beans, peas and lentils are the only vegetables with an appreciable level of protein. However although a 4-oz portion of beans contains just as much protein as an egg, not all that protein can be used by the body. Protein foods are made up of a variety of amino acids, eight of which cannot be manufactured by the body and so have to be eaten in food. To be 100 percent usable by the body, the eight amino acids have to be present in food in the correct proportions.

The problem with bean protein is that it is over-endowed with an amino acid called lysine and defficient in one called methionine. This problem can be overcome if beans are eaten with another protein food, such as cereals, which are rich in methionine and weak in lysine. Another complementary protein mix is beans and seeds or nuts (but remember that peanuts are legumes too), and beans and dairy produce. To get the very best out of these complementary mixes, use about 1·2 parts beans to 1 part cereals, 1/3 beans to 1/2 seeds, and use a little fresh or powdered milk in all bean dishes. Baked Beans on Toast, Beans with Rice, Hopping John, Succotash, Dahl and Rice, and Hummous all make use of complementary proteins.

SOYA BEANS AND TEXTURED VEGETABLE PROTEIN

Soya protein is of a much better quality than that of other beans, and considerably more of it can be used by the body even if it is eaten on its own. The harvest from one acre of ground planted with soya beans can produce enough protein food for one person for 6–7 years. Compare this to the same harvest fed to animals for slaughter. The meat provides only enough protein food for that one person for 77 days.

The research into soya bean protein led first to the development of textured vegetable protein or TVP. TVP is produced by extracting 70% protein from the bean, leaving behind most of the carbohydrate and fat. It can be used as a food in its own right, or as a meat extender in hamburgers, canned foods, pies and fishcakes. The products tended to have a strong bean taste and smell, and the flavoured items did not really taste like meat. However, the food industry took up TVP and used it fairly extensively.

More recent research has led to much more sophisticated products with a protein concentration of as much as 95–99%, and without the problems associated with TVP. These soy protein isolates can be used in powder form in almost any kind of food, and they can also be spun into edible fibres. They are basically food manufacturers' products.

GROWING BEANS

Most beans prefer temperate climates, but broad beans will tolerate a slight frost, and the long roots of the soya bean make it very resistant to drought. Beans are relatively easy to grow. A limed soil will increase yield, and a shallow soil will need manure.

Soya beans will grow in almost any kind of soil and, when grown in ideal conditions, will mature in September or October. The mature plant loses its leaves and the seeds dry on the plant to a level at which they can safely be stored. The crop is then harvested, either by hand or with a combine harvester.

In Britain and the United states broad beans, French beans and runner beans (Snap beans, stringless beans and greenbeans) are all grown from seed. Some green beans are grown for the fresh vegetable market, but very many more are grown for canning factories and frozen food manufacturers. The manufacturing units take an active part in the growing process and work very closely with farmers under contract to them to ensure a high standard of quality and consistency. Teams of experts go into the fields to determine exactly when the beans should be picked. The harvesters move in and the crop is often in the freezer or the can before the end of the day.

On St Valentine's Day, cast beans in clay,
But on St Chad's Day sow good or bad.

	SOWING	CARE
Broad beans	Sow in late autumn, February or March; sow every 2–3 weeks for successive crops Sow in double row in shallow trench 2" deep, 9" apart; 2' between rows.	Erect central support. Pinch out tips of plants when flowers appear to encourage growth. Watch out for black fly, mice and bean beetle.
Runner beans	Sow out in sheltered position in rich soil once risk of frost has passed. Or sow indoors in mid-Spring and plant out in late-Spring. Sow in double row 2" deep, 6" apart. 1' between rows. Thin to 1' apart.	Support with long poles and string. Pinch out growing tips when plants reach top of poles. Water well. Watch for aphids and slugs.
Dwarf or French beans	Sow in mid-Spring in well dug soil. Sow every 2–3 weeks for continuous crop. Sow 1–2" deep, 4–6" apart. 1' between rows.	Water regularly.
Haricot beans	As Dwarf French beans.	

HARVESTING	PATIOS AND WINDOW BOXES
Harvest regularly. Pick when young and tender and eat pods, as *mange tout*. Plants grow to 4–5' (dwarf varieties to 15".)	Grow in small patio beds in well drained soil. Grow from small plants. Water well.
Harvest regularly to encourage growth. (A 10' row can yield over 50 lbs.) Plants are climbers and will grow to 6–9'.	Like a sheltered sunny wall. Manure a narrow bed well. Support with plastic netting, poles and string. Use wigwam shape for small patio beds.
Eat when small and thin. Plants are leafy so search for bean pods.	Sow 4" apart diagonally across box, 1" deep. Soak daily with tepid water until shoots show. Water regularly, make sure soil well drained.
Leave until pods are dried and brown. Lift and hang in an airy shed or garage to dry.	

SPROUTING BEANS

T he Chinese have been sprouting beans for thousands of years but the idea is relatively new to the West. Bean sprouts have a delicious flavour and a lovely crunchy texture. They are also very nutritious. Added to the original protein content of the bean are a host of vitamins and minerals which appear as the bean starts to sprout and change its form. Vitamin B complex and Vitamin C both show enormous increases, as do Vitamins A, E and K. In addition, sprouts are an excellent source of enzymes which help in the digestion process; they contain minerals in a form easily used by our bodies, and they also contain natural fibre.

Bean sprouts are one of the freshest vegetables we have since they are still growing when we eat them. This means that we get all the nutritional value before it has had a chance to dissipate.

STEP-BY-STEP GUIDE TO SPROUTING BEANS

Sprouting must be one of the easiest methods of cultivation. All you need is a large jar or container, a piece of muslin or gauze to fit over the opening and an elastic band to keep it in place. Alternatively you could use a colander lined and covered with muslin. Or, if you plan to do a lot of sprouting, you could invest in a purpose-made sprouter.

1. Wash the chosen beans and pick out any that are damaged or broken.

2. Place 2 tablespoons of beans in your container ($\frac{1}{2}$ oz beans can become 2–4 oz of sprouts!).

3. Some beans need to be soaked first. If so, cover with at least four times their volume of water to allow for expansion. Leave to stand overnight. In the morning drain well and leave to stand at normal room temperature. (Place jars on their sides, slightly tipped, so that any excess water can drain away.)

4. Twice or three times a day (see guide) rinse the beans with cold water and drain off. Repeat daily for 3–5 days until the sprouts have reached the desired length. In cold weather they will grow more slowly, so help them along by using tepid water. Bean sprouts that are not rinsed can go mouldy.

BEWARE!

★ *Do not* place the container over direct heat or in direct sunlight – it will not improve the growth rate, rather do the opposite. A well shaded area is best.

★ *Do not* rinse with hot water – they may start to rot.

★ *Do not* allow the beans to stand in a cold draught.

★ *Discard* any beans which have not sprouted after six days, and any sprouts which have gone mouldy. (Be careful not to confuse mould with the tiny furry roots that some beans naturally grow.)

GUIDE TO BEANS FOR SPROUTING

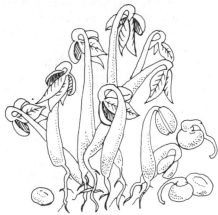

The popular Chinese bean sprouts are grown from the mung bean, but other beans can be successfully sprouted. Always buy beans from a shop which does a brisk business in beans or they may be too old to sprout.

MUNG BEANS: These small green beans increase by a factor of four and will be ready in 3–5 days. Mung bean sprouts grown at home are greener than commercially grown ones, which are grown in the dark.

Rinse: 2–3 times daily.
Soak: No need in a sprouter, otherwise soak overnight.

ADUKI BEANS: These small dark red beans are sometimes referred to as red soya beans, but they are smaller than soya beans and not quite as difficult to sprout. They taste similar to mung beans and can be substituted for them in recipes. They usually increase by a factor of 4 in 4–5 days.

Rinse: 3–4 times daily.
Soak: No need in a sprouter, otherwise soak overnight.

CHICK PEAS/GARBANZO BEANS: These round yellow beans taste just like freshly podded peas when they have been sprouted. They double in volume and reach a good length in 2–3 days. Do not let them grow too large or they may be a little tough.

Rinse: 2–3 times daily.
Soak: No need in a sprouter, though they may take a little longer to sprout, otherwise soak for 12 hours.

SOYA BEANS: These can be fairly difficult to sprout, once sprouted should be steamed or boiled for a few minutes before eating to remove the trypsin inhibitor which can make them difficult to digest.

Rinse: 4–6 times daily.
Soak: 15 hours.

USING BEAN SPROUTS

Bean sprouts are at their best straight from the sprouter, but keep in the fridge for up to a week.

The full value of bean sprouts, with the exception of soya sprouts, is probably gained by eating them raw. They can be added to a salad or they can be mixed with cold scrambled eggs, grated or cream cheese, or peanut butter for a crunchy sandwich filler.

If you do decide to cook bean shoots, take a tip from the Chinese and keep cooking to a minimum.

STIR-FRIED BEAN SPROUTS

2 tablespoons cooking oil
1 small onion, sliced
1 inch fresh root ginger, cut into thin sticks
2 carrots, cut into thin sticks
6 oz bean sprouts (any kind)
2 tablespoons soy sauce
black pepper

Heat the oil in a frying pan or wok and fry the onion, ginger and carrots for 2 minutes, stirring and tossing all the time. Add the bean sprouts and cook for a further minute. Add the soy sauce and pepper. Toss to heat through and serve at once.

Serves 4 as a vegetable

BEAN SPROUT AND PASTA SALAD

*4 oz pasta spirals, tubes or butterflies, cooked, drained
and mixed with*
1 tablespoon salad oil
4 oz mung bean sprouts
2 oz button mushrooms, sliced
1 small green pepper, seeded and thinly sliced
3 sticks celery, sliced
2 oz salted peanuts
¼ pint yoghurt
2 tablespoons peanut butter
1 tablespoon lemon juice
milk

Leave pasta to cool, then mix in the vegetables and
peanuts. Mix all the remaining ingredients with
sufficient milk to make a smooth creamy dressing and
pour over the salad.

BEAN SPROUT OMELETTE

Fry 4 oz streaky bacon to release some of the fat, then
add 6 chopped spring onions. Continue frying until the
onions are tender and the bacon cooked through. Add
4 oz aduki bean sprouts, herbs and seasoning and toss
well together to heat the bean sprouts. Remove from
the heat and keep on one side.

Make one large or four small omelettes. Just before
folding over, add the bean sprout mixture.

CHOOSING AND
USING FRESH GREEN
BEANS

Whatever the bean, buy or pick them as young as possible. Always eat directly after picking or buying. If absolutely necessary, store in a cool, dry place. Broad beans can be podded and stored in a rigid container in the fridge.

Serve beans cooked in main dishes, as a vegetable accompaniment or in salads with a vinaigrette. Really young beans can also be eaten raw in salads.

BROAD BEANS (WINDSOR OR SHELL BEANS)

Shopping: Fresh broad beans are available from the end of May to early September with a peak in June and July. Choose young beans with fairly soft tender pods.

Preparation: Top and tail very young beans and cut into lengths, pods and all. Steam in a very little water and serve with parsley sauce. Pod older beans and boil in salted water. Cook as soon after podding as possible. Serve tossed in butter or cream or with parsley sauce.

As the season progresses the beans tend to develop a tough outer skin which can be removed by plunging the raw beans into boiling water for a few minutes and rubbing of the skins. The beans are then returned to the pan to continue cooking. Alternatively, they can be sieved after cooking.

Cook with a sprig or two of parsley or savory in the cooking water to bring out the flavour, and serve with duck, goose, pork and bacon.

FRENCH BEANS (DWARF, KIDNEY, STRINGLESS BEANS)

Shopping: Fresh French beans are available all the year round but they can be quite expensive. Home-grown varieties are in the shops from early June until September. The beans should be firm and fresh and juicy when cut or snapped between the fingers.

Preparation: These beans usually need only to be topped and tailed. Leave whole, or cut larger beans into lengths. Very old beans may need to be stringed.

Cook the beans in lightly salted boiling water until just tender. Flavour with sprigs of savory or basil. Drain thoroughly and serve tossed in butter or cream or serve cold with vinaigrette.

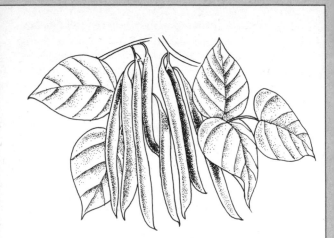

RUNNER BEANS

Shopping: Fresh runner beans come into the shops in mid-July and last until the first frosts, but they are grown under glass for out of season sales. The pods should be green and succulent with the developing seed small and juicy. The bean should not be limp but should break cleanly. Avoid those which are badly curled or old and stringy.

Preparation: Apart from very young beans, runner beans need to be stringed and sliced.

Cook the sliced beans in lightly salted boiling water until just tender. Drain thoroughly and serve tossed in butter or cold with vinaigrette.

MIDDLE EASTERN LAMB
WITH BEANS

Beans with mutton or lamb is an international combination which is just as likely to turn up in France or Britain as it is in the Middle East or in India. This recipe from the Levant uses French beans, but others use dried haricot or kidney beans.

1 lb lean lamb, cut into chunks
2 large onions, sliced
3 oz butter
1 lb French beans (fresh or frozen and thawed)
1½ tablespoons tomato paste
1 pint water
salt and pepper

Fry the meat and onions in butter for 2–3 minutes. Add the beans and simmer gently for 10 minutes, shaking the pan from time to time.

Stir in the tomato paste and then pour on the water. Season, cover and bring to the boil. Simmer very gently for 3 hours.

Serves 4

Sow four beans in a row,
One for cowscot, one for crow;
One to rot and one to grow.

BAKED GAMMON AND BEANS

This recipe is typical of the traditional English Beans and Collops served at 16th and 17th century feasts. It may even be the 'Leicester beans and bacon – food of kings' referred to in William King's *Art of Cooking*. In later years a roux sauce may have been made to carry the parsley. If you prefer this method, serve with parsley sauce poured over the top.

1¼ lbs sliced gammon or a small bacon joint
1¼ pints red wine or half wine and water
1 oz brown sugar
3 cloves
1 lb fresh or frozen broad beans
1 oz butter
3 tablespoons freshly chopped parsley
½ teaspoon dried savory

Arrange the gammon slices in a baking dish. Mix the wine, sugar and cloves and pour over the top. The liquid should just cover the meat. Bake at 350°F/ 180°C/ Gas 4 for 1 hour. If using a joint use the same amount of liquid and baste frequently during cooking; it may need to be cooked for longer.

Cook the beans in lightly salted boiling water and toss in butter with the herbs. Pile the beans in the centre of a serving plate and arrange the slices of meat around the outside.

Serves 4

FRENCH BEANS LYONS STYLE

The French have quite a number of different ways of dealing with fresh *haricot verts*. This recipe stems from the Lyons area of the Rhône valley.

1 lb French beans (fresh or frozen and thawed)
1 small onion, very finely chopped
1 oz butter
salt and pepper
2 tablespoons wine vinegar or dry white wine
1 tablespoon freshly chopped parsley

Top, tail and string fresh beans and pick over the frozen ones.

Gently fry the onion in half the butter until soft but not brown. Season and add the beans and the rest of the butter. Continue to cook until the beans begin to change colour.

Add the vinegar or wine and finely chopped parsley. Simmer for a further minute or so and serve in a warmed serving dish.

Serves 4

Be it weal or be it woe,
Beans blow before May doth go.

BEAN STUFFED TOMATOES

Puréed broad beans may seem a little odd to modern tastes, but in the past cooks puréed all but the smallest and youngest beans. One recipe from Mrs Glasse's *Cookbook* suggests layering puréed and whole beans and baking in a casserole. This recipe is a little more modern and looks attractive with roast meats or grills.

8 oz broad beans (weight after podding, or frozen)
1 teaspoon dried savory
½ oz butter
salt and pepper
4 large tomatoes, halved
1 oz breadcrumbs

Cook the beans as directed on the pack, but with a teaspoon of dried savory in the water. Drain well and sieve or purée in a blender. Stir in the butter and season to taste.

Scoop out the pulp from the tomato halves and fill each one with a little of the bean purée, and sprinkle with breadcrumbs. Place on an ovenproof dish. Bake at 400°F/200°C/Gas 6 for 25 minutes.

Serves 4

CHOOSING AND USING DRIED BEANS

Dried beans will not keep in top condition for ever, and the older they are the harder they will be when cooked. Always buy beans from a shop that has a high turnover, and never buy too many at once. Store in airtight jars in the dark. Very occasionally you may come across a bean weavil; if so, throw out the beans and start afresh.

All dried beans need to be soaked before use. Cover with plenty of cold water and leave to stand overnight. If you are in a hurry, pour boiling water over the beans and leave them to stand for 2 hours. To test if the beans have been soaking for long enough, break one in half – it should be the same colour all the way through. If bubbles have collected in the soaking water the beans have been there too long, but they can still be used.

Always drain the soaking water from the beans, and remove any beans which have not taken up the water. Cook in fresh water. Do not add any salt at this stage as it tends to toughen the beans. Season later. Cook in boiling water until tender or, for a much faster cooking time, use a pressure cooker.

Some experts advocate adding a pinch of baking soda to the cooking water to help soften the beans, others add it to the soaking water. But in either case it should be used sparingly as it can affect both the flavour and the nutritional value.

Cooking times for different beans are given below but they can only be approximate as the length of time a bean takes to cook will depend upon how old it is, how hard the water is and how long it has been soaked. Test the beans with a fork; overcooked beans tend to split and lose their shape.

If you are in a hurry use canned beans, but remember to compensate for the fact that the quantity of dried uncooked beans will be rather more after soaking and cooking. Beans can take up as much as their own weight in water. Canned beans are already cooked.

ADUKI BEANS: Small dark red beans with a nutty sweet taste, used in sweet pies and purées as well as in savoury dishes. Cooking time: 30 minutes on the hob or 10 minutes in the pressure cooker.

BLACK BEANS: Black shiny beans looking a little like large haricot beans. They add interest to bean mixtures. Cooking time: $1\frac{1}{4}$ hours on the hob or 20 minutes in the pressure cooker.

BLACK-EYE BEANS (peas): White rounded beans with a black mark like an eye at one side, traditional in Creole dishes. Cooking time: 1 hour on the hob or 20 minutes in the pressure cooker.

BORLOTTI BEANS (ROSE COCO AND CRANBERRY BEAN): Quite large pink beans with a speckled coat, very popular in Italy. They have a slightly sweet taste.

Cooking time: 1 hour on the hob or 20 minutes in the pressure cooker.

BROAD BEANS (FAVE BEANS): Large, flat, brown beans, traditionally served in winter but fairly difficult to find. Cooking time: 1½ hours on the hob or 30 minutes in the pressure cooker.

BUTTER BEANS (MADAGASCAR BEANS): Very large, flat, white beans, popular in Britain. Available canned. Cooking time: 1½ hours on the hob or 30 minutes in the pressure cooker.

FOULE MEDAME (FOULIE BEANS): Round brown beans with a very thick skin, popular in the Middle East. Cooking time: 1¼ hours on the hob or 20 minutes in the pressure cooker.

GARBANZO BEANS/CHICK PEAS Medium-sized yellow beans which are round in shape and slightly wrinkled. Available canned. Cooking time: 1¼ hours on the hob or 25 minutes in the pressure cooker.

HARICOT BEANS: Small, white, oval beans which are the dried seed of the French or kidney bean. It has a lot of very close relatives – one of these is the French Flageolet bean, a pale green bean which is longer and thinner than the haricot; another is the Italian Canellini (fazolia) bean, which is white but larger than the real haricot bean. Available canned. Cooking time: 1¼ hours on the hob or 20 minutes in the pressure cooker.

MUNG BEANS: Small, round, green beans more often used for sprouting than cooking, but they can be soaked and cooked. Cooking time: 30 minutes on the hob or 10 minutes in the pressure cooker.

NAVY BEANS: Small white beans usually used to make baked beans; many different kinds of dried white beans come under this name. Cooking time: 1 hour on the hob or 20 minutes in the pressure cooker.

PINTO BEANS: Dappled pink medium-sized beans which are the frijoles of Mexico. Cooking time: 1 hour on the hob or 20 minutes in the pressure cooker.

RED KIDNEY BEANS: Red, shiny beans which are kidney shaped. These beans contain a substance which can be dangerous if they are not cooked properly. Always fast boil for at least 10 minutes before proceeding with any cooking method. Available canned. Cooking time: 10 minutes fast boiling plus 1 hour on the hob or 20 minutes in the pressure cooker.

SOYA BEANS: Small, yellow beans which are oval in shape. They should be soaked for at least 15 hours in the fridge and must be well cooked to make them digestible. However long they are cooked they always retain a slight 'bite'. Available canned. Cooking time: 2 hours on the hob or 30 minutes in the pressure cooker.

HOUMMOUS

This creamy purée of chick peas and sesame seeds turns up on most Middle Eastern Mese selections. Serve with hot pitta bread.

8 oz chick peas or garbanzo beans, soaked overnight in
*   cold water and drained*
1 clove garlic, crushed
salt
4 oz sesame seeds, ground
3 fl oz lemon juice
2 fl oz salad oil
3–4 black olives
sprigs continental parsley

Cook the chick peas in plenty of boiling water for about 1¼ hours until tender. Drain well and rub through a sieve.

Stir in the garlic, salt, sesame seeds and lemon juice. Mix to a smooth, almost runny, paste. Next swirl in the oil.

Serve garnished with black olives and parsley.

Purées of dried beans used to be the ploughman's staple food. Boorde recalls 'Beene butter is use moche in Lent in dyvers countries – it is good for plowmen to put in their paunches.' Bean butter seems to have been made from dried winter beans boiled to a mush with mutton broth and used as a thick spread upon coarse oatcake.

BEAN AND GREEN VEGETABLE SOUP

This recipe is based on the French Garbure and the variations were inspired by a Caribbean version of the same soup from Martinique. Serve as a main course soup with large chunks of wholemeal bread.

1 onion, finely chopped
1 clove garlic, chopped (optional)
3 oz streaky bacon, diced
1 tablespoon cooking oil
1 large potato, diced
2 carrots, diced
4 oz large haricot or cannellini beans, soaked overnight in cold water and drained
8 oz split beef bones
2½ pints water
2-3 sprigs fresh thyme or ½ teaspoon dried thyme
½ teaspoon dried marjoram
salt and pepper
1 lb savoy or white cabbage, cut into slices
or
8 oz cabbage with 1 green pepper, seeded and chopped and 4 oz okra
1 green chilli, seeded and finely chopped
2 hard-boiled eggs, chopped (optional)

Fry the onion, garlic (if used) and bacon in the cooking oil until lightly browned. Add the potato, carrots and beans and toss in the fat for a minute or so. Next add the bones, water and herbs. Bring to the boil, cover and simmer for 1 hour.

Season and add the remaining vegetables. Cook for 15–20 minutes until the cabbage is just tender.

Serve sprinkled with chopped hard-boiled eggs if desired, and serve with rye or wholemeal bread.

Serves 4

SUCCOTASH

This is a traditional American dish which uses cooked Lima beans. Buy canned vegetables for an easy accompaniment to roast turkey, beef or lamb.

1 oz streaky bacon, diced
1 dessertspoon plain flour
¼ pint single cream
4 oz cooked turnip or swede (optional)
1-lb can Lima beans or flageolet, drained
7-oz can sweetcorn kernels, drained
salt and pepper

Gently fry the bacon in the pan to release the fat. Stir in the flour and then the cream. Bring to the boil. Add all the remaining ingredients. Stir and return to the boil and simmer for 5 minutes.

Serve in a warmed dish.

Serves 4–6

GREEK BEAN SALAD

In Greece this salad is made with small round brown beans which are locally grown, but it is just as good with rose coco or Borlotti beans.

4 oz dried Greek Foulia beans, rose coco or borlotti beans, soaked in cold water overnight and drained
1 small onion, thinly sliced
4 tomatoes, cut into quarters
3 tablespoons olive oil
1 tablespoon wine vinegar
½ teaspoon dried marjoram
¼ teaspoon dried thyme
salt and pepper
2 oz Feta cheese, crumbled
6–8 black olives
2 tablespoons freshly chopped parsley

Cook the beans in fresh gently boiling water until tender – about 1¼ hours, or in the pressure cooker for 20 minutes. Take care that the beans do not split. Drain and leave to cool.

Mix with onion and tomatoes and spoon into a serving dish.

Mix the oil, vinegar, herbs and seasoning and pour over the salad. Crumble Feta cheese over the top and decorate with olives and parsley

Serves 4

BEAN POT

There must be hundreds of Bean Pot recipes and it is very easy to make up your own personal version if you vary the type of beans used, the flavourings and the meats. Incidently, this version also makes a very good vegetarian dish: simply omit the bacon and serve with hunks of wholemeal bread.

2 tablespoons cooking oil
4 oz streaky bacon, diced
2 onions, finely chopped
1 clove garlic, finely chopped
2 sticks celery, finely chopped
3 oz haricot beans, 3 oz barlotti beans, and
2 oz butter beans, soaked overnight in cold water and
    drained
2 tomatoes, peeled, seeded and chopped
pinch fennel seeds
pinch mixed herbs
salt and pepper
½ pint red wine

Heat the oil in a frying pan and fry the diced bacon for 2–3 minutes. Add the onion, garlic and celery and continue frying until the vegetables are softened.

Stir in all the remaining ingredients except the wine and spoon into a small deep casserole. Add the wine, cover and bake at 350°F/180°C/Gas 4 for about 2 hours until the beans are tender.

Serves 4

CASSOULET

This classic dish from the Languedoc of Southern France is really the French version of Bean Pot. It is basically a stew of white beans with different meats, and recipes vary from district to district.

1 tablespoon dripping or goose or chicken fat
2 onions, sliced
4 oz streaky bacon, chopped
1 lb belly of pork, cut into slices
8 oz smoked, garlic or ordinary pork sausages, sliced or
  cubed
8 oz butter beans or half butter beans and half haricot
  beans, soaked overnight in cold water and drained
3 tomatoes, skinned and chopped
1 bay leaf and 1 bouquet garni
salt and black pepper
1 pint water
2 oz fresh breadcrumbs

Melt the dripping in a frying pan and fry the onion for 2–3 minutes. Add the bacon, belly of pork and sausages and fry until they are all well browned.

Layer in a casserole with the beans, tomatoes, herbs and seasoning. Pour on the wine.

Cover and bake at 325°F/170°C/Gas 3 for 2 hours. Remove the lid. Sprinkle with breadcrumbs and continue cooking for $\frac{1}{2}$–$\frac{3}{4}$ hour until the beans are tender and all the liquid has been taken up.

Serves 4

URHAD DAL

Urhad are small black beans used widely in India. They can be bought split or whole from Indian grocers. However, the same recipe can be made using whole Mung beans or even haricot beans.

4 oz whole urhad, mung or haricot beans, soaked overnight in cold water and drained
1 teaspoon turmeric
1 inch fresh root ginger, finely chopped
1 onion
1 tablespoon cooking oil
2 tablespoons freshly chopped parsley
1½ tablespoons garam masala
1½ tablespoons ground cumin
chilli powder to taste
2 tomatoes, peeled and chopped
4 tablespoons water
2 teaspoons lemon juice

Cook the beans in boiling water with the turmeric for ¾ hour.

Fry the ginger and onion in cooking oil until the onion begins to soften. Add the parsley and spices and stir well. Add the tomatoes and cook until the mixture begins to sizzle.

Drain the beans and add to the mixture with the water. Continue cooking for a further 30 minutes.

Add the lemon juice and serve with Indian bread or rice.

Serves 4

MEXICAN REFRIED BEANS
WITH SPINACH

Refried beans are popular all over South America. They are beans which have been boiled and then fried in bacon or pork fat – and they can be just a little bit boring. This recipe comes from New Mexico and it gives much more flavour to the refried beans.

4 oz streaky bacon, diced
2 onions, finely chopped
1 clove garlic, crushed
8 oz cooked pinto beans or red kidney beans
1 fresh chilli, seeded and chopped
½ teaspoon cinnamon
1 lb frozen spinach, cooked as directed on the pack
salt and pepper

Fry the bacon until crisp. Remove the pieces from the bacon dripping and add the onion and garlic. Fry until lightly browned. Add the beans and continue to fry for 5–6 minutes, stirring and crushing about half the beans.

Stir the chilli and cinnamon into the cooked spinach and season to taste. Spoon into a serving dish, and top with refried beans.

Serves 4

Preserving Beans

Freezing Beans

Pick or buy all beans for freezing when they are still fairly young. Shell or string and slice and blanch before freezing by plunging first into boiling water and then into ice-cold water. Pack into bags or open freeze and then pack. Store for up to 12 months.

Salting Beans

Our great grandmother would have salted or pickled beans, and it is still worth doing if you have a glut of runner beans.

Choose young beans and allow 1 lb kitchen salt to each 3 lbs beans. (Do not use table salt as this has a chemical added to help it run freely which affects its preserving properties.) Top, tail and string the beans and cut into lengths. Put a thick layer of salt in the bottom of a preserving jar, and then put alternate layers of beans and salt, ending with salt. Press down thoroughly so that there are no air holes.

Cover the jar and leave for four days, then add more beans and a little more salt to fill the jar. Cover with a moisture-proof lid. The salt will draw the water from the vegetables to make a thick brine and the beans must be completely covered by the brine. If at any time they are not, add more salt. Store in a cool dry place for up to a year.

To use salted beans, soak in several batches of fresh cold water. It will take 4–5 hours to remove all the salt. Cook in boiling water with no added salt until tender.

PICKLED BEANS

Pickled green beans are popular in America where there are a number of traditional recipes.

2 lbs young green beans, topped and tailed and stringed if necessary
2 cloves garlic
4 sprigs fresh dill weed or 4 teaspoons dill seeds
3 teaspoons salt
18 fl oz vinegar
9 fl oz water

Cut the beans into lengths if they are very long, and cook in salted boiling water for 6–8 minutes until just tender. Pack into a kilner jar so that they stand upright, and add 2 cloves of garlic, dill and salt. Mix the vinegar and water, pour over beans to fill jar.

Keep the jar in the fridge for up to 2 weeks. Serve the beans drained.

ENGLISH BEAN CHUTNEY

If you have a glut of runner beans try this 19th-century recipe for chutney.

$1\frac{1}{2}$ *lbs onions, chopped*
$1\frac{1}{2}$ *pints vinegar*
2 lbs runner beans, stringed and cut into 1–inch lengths
2 lbs soft brown sugar
1 oz cornflour
$\frac{3}{4}$ *oz turmeric*
$\frac{3}{4}$ *oz dry mustard*

Place the onions in a pan with 1 pint vinegar and bring to the boil. Simmer for 30 minutes and add the beans. Cook for a further 10 minutes, remove from the heat and stir in the sugar.

Mix the cornflour, turmeric and mustard to a smooth paste with the rest of the vinegar and pour into the pan. Bring to the boil, stirring all the time. Continue cooking until the mixture is reduced to a thick consistency.

Pot and seal in jars.

Makes approximately 5 lbs

DRYING BEANS

It is not often worth drying beans at home but if you do want to have a go here's how.

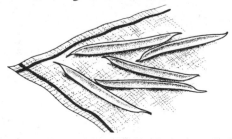

FRENCH AND RUNNER BEANS: Wash, dry, top and tail the beans and string if necessary. Slice fairly thickly or leave small French beans whole. Blanch quickly in boiling water and refresh in cold water. Dry on kitchen towelling. Spread out on trays and dry in a very cool oven 140°F/60°C. Cool and pack into jars to store. Soak in water before using.

HARICOT BEANS: Follow the procedure set out on page 17, and when the pods are light yellow brown in colour and parchment dry, shell the beans and spread out on pieces of brown paper to finish drying out. When dry, store in lidded containers. Soak before using.

'Bean pods are noisiest when dry
And you always wink with your weakest eye'.

Bret Harte: *The Tale of a Pony*

BAKED BEANS

More baked beans are eaten in Britain than in any other country in the world. A massive two million cans are brought every day! Yet when baked beans were first introduced they were rather expensive and not very popular. The London store, Fortnum and Mason, was the first stockist and the beans cost about 3½p per 16–oz can – around 3% of the average worker's weekly wage.

Baked beans are made from navy beans, which are grown extensively in the United States, and the recipe is based on traditional New England recipes for Baked Bean Pot. In the days of the Puritans the Sabbath lasted from sundown Saturday night to sundown the following evening, and the Bean Pot provided a hearty dish which was easy to prepare and left time for the strict Sabbath observancies.

New England was also the place where canned baked beans first appeared. This was in 1875 when the Portland, Maine, firm of Burnham and Morrill supplied their fishing fleet with cans of 'Boston-type' baked beans flavoured with pork and molasses. Twenty years later, in 1895, the Pittsburg firm of H. J. Heinz produced baked beans in tomato sauce. The cans each contained a piece of pork as a flavouring ingredient, and this practice continued through the growth of H. J. Heinz in Britain until wartime shortages necessitated the introduction of vegetarian baked beans, such as those on sale in the shops today.

BOSTON BAKED BEANS

For the traditionally minded – with time on their hands – here's a recipe for making your own old-style baked beans at home.

1 lb navy or small haricot beans, soaked overnight in
*   cold water and drained*
1 pint water
1 teaspoon salt
1 oz soft brown sugar
1 teaspoon salt
1 oz soft brown sugar
1 teaspoon dry mustard
2 fl oz molasses (approximately 4 tablespoons)
4 oz salt pork, cut into chunks
1 onion, sliced

Cover the beans with the water, bring to the boil and simmer for 1 hour. Drain the beans and mix the liquid with the salt, sugar, mustard and molasses.

Place the beans in an earthenware pot with the pork and onion and pour the liquid over the top. Cover and bake at 300°F/150°C/Gas 2 for 4–5 hours. Check from time to time to make sure that the beans are not drying up. If they are, add a little more water.

Serves 4

CURRIED BEAN AND APPLE SOUP

1 oz butter
1 onion, chopped
1 lb cooking apples, cored and chopped
2 x 8-oz cans curried beans with sultanas
1 pint water
1 chicken stock cube

Melt the butter and fry the onion until transparent. Add all the remaining ingredients. Stir and bring to the boil. Cover and simmer for 20 minutes. Sieve or purée in a blender and reheat. Check seasoning.

Serves 4

BAKED BEAN CASSEROLE

1½ lbs spare rib pork, boned and cubed
1 tablespoon cooking oil
1 lb swede, peeled and cubed
1 lb leeks, sliced
8 fl oz dry cider
16-oz can baked beans

Fry the pork in cooking oil until well sealed. Mix in all the remaining ingredients and spoon into a casserole. Cover and bake at 375°F/190°C/Gas 5 for 1–1¼ hours until the meat is tender. Serve with jacket potatoes baked at the same time.

Serves 4

BEANS AND THE ORIENT

FERMENTED BEAN PRODUCTS

Soy Sauce is by far the best known of the fermented bean products, and this salty, dark brown liquid makes an excellent seasoning for Eastern-style food. Indeed, this was probably the basis for its development in China, a country with virtually no rock salt.

In Japan soy sauce is known as shoyu and is used in much the same way as in China. Tamari is another fermented bean product made from soya beans.

Some modern soy sauces are made with fermentation from hydrolyzed vegetable protein, caramel colouring, corn syrup, salt and pepper. These may look and taste like soy sauce but they do not have the same power to enhance the savouriness of the food. *Miso* is another savoury fermented seasoning made from soy beans, rice or barley, salt and water and a special kind of mould. It was originally made in China but has become very much a Japanese classic.

Its texture resembles that of peanut butter and it varies in colour from yellow to deep reddish brown. It is pretty strong tasting and should be used sparingly. Use it like a stock cube in soups and stews or as a flavouring for dips.

Yellow and black bean sauces In China these two sauces are used to flavour stir-fried food or as dipping sauces on the table. Yellow bean sauce is made from the concentrated residue which remains after the natural fermentation of soy sauce, and it is rich in vitamins. It is a very powerful flavouring. Black bean sauce is also made from fermented beans; the beans may be mixed with garlic or with chillis.

CHICKEN IN YELLOW BEAN SAUCE

1 lb boned and skinned chicken meat
3 tablespoons cornflour
4 tablespoons cooking oil
2 oz cashew nuts or unsalted peanuts
1 jar Peking Yellow Bean Stir-Fry Sauce

Cut the chicken into small pieces and toss in cornflour. Heat the oil in a wok or frying pan and fry the nuts. When the nuts start to change colour, add the chicken and stir-fry for about 2 minutes until there is no trace of pinkness in the chicken. Add the sauce and heat through quickly. Serve at once with rice and stir-fried beansprouts.

SOY MILK AND BEAN CURD OR TOFU

In Japan the making of soy milk and tofu has reached a high art, and people are as likely to talk of the relative merits of different tofu shops as, in the West, we are to compare the wine from different vineyards.

Soy milk is made by soaking whole uncooked soya beans overnight and then grinding the beans to a thick paste. This paste is cooked quickly in boiling water and the resultant mixture is separated into solids, 'okara', and liquid, 'milk'.
Soy milk can be bought canned or dried and it makes an excellent replacement for baby milk or a substitute for children and adults who are allergic to cow's milk.

Bean curd or tofu is made by using lemon juice, vinegar, calcium or magnesium chloride to curdle soy milk in much the same way as rennet is used to curdle dairy milk. The soy milk separates into curds and whey. The whey is drained off and the curds pressed to give a firm consistency. Chinese bean curd is usually pressed into a much more compact product than Japanese tofu.

Silken tofu is made from very thick soy milk and the curds and whey are never separated, nor is the tofu pressed. It is more difficult to use in cooking, though being softer it does purée more easily.

Tofu is an extremely healthy food. It does not contain any cholesterol and is high in polyunsaturated fats. It is high in protein but low in calories, containing only 56 calories per 4 oz portion. Interestingly, it can contain as much calcium as dairy milk, as well as significant quantities of iron, phosphorous, potassium, sodium and vitamins B and E. Silken tofu retains more of the natural goodness of the soy milk than regular tofu since the whey has not been removed. Both types can be found in most health food shops.

DEEP FRIED SPICY TOFU BITES

8 oz fresh tofu
3 tablespoons plain flour
1 tablespoon garam masala or mixed curry spices
1 egg, beaten
2 oz dry breadcrumbs
oil

Cut the tofu into bite-sized cubes and place between layers of kitchen paper and leave to stand for 15 minutes to reduce the moisture content.

Mix the flour with the garam masala or curry spices and toss the tofu in the mixture. Dip each piece into beaten egg and coat well with dried breadcrumbs. Heat the oil and deep fry for 2–3 minutes turning during cooking.

Serve with wedges of lemon and soy sauce.

Makes 16–20

BEANS AND MEDICINE

Beans do not appear too often in medical tracts, except perhaps as a cause of flatulence! Nor do they figure in many herbals. However, there are quite a few experts today who believe in the health-promoting properties of bean sprouts.

In Japan, okara, the residue from making soy milk, is reputed to be both a cure for diarrhoea and a help to nursing mothers who take it to increase their supply of breast milk. Wrapped in a cloth it is also said to act as an excellent wax polish to help with the housework!

OLD WIVES' CURE

In the West, the only reference to the medicinal properties of beans seems to have been in the old wives' cure for persistent coughs – beans boiled with garlic.

FLATULENCE

The main complaint against dried beans is that they cause flatulence. The Elizabethan herbalist Gerard advocated the use of 'the fruits and pods of kidney beans boiled together before they be ripe, these are an excellent delicate meat and do not engender winds as other pulses do.'

Beans contain oligosaccharides which cause flatulence if they are not broken down by soaking, draining and proper cooking in fresh, unsalted water. If they are eaten they are attacked by bacteria in the gut and this produces the intestinal gas. The addition of herbs like Summer Savory, and particularly Sweet Cicily, can help to reduce the flatulence problem. Enzymes at work in bean sprouts break down the oligosaccharides before they are eaten.

RAW BEANS

Dried beans should never be eaten raw. Uncooked beans can cause headaches, and pains and irritation in the gut, and some raw beans such as red kidney beans can kill.

BAN ON BEANS

Many writers have been suspicious of beans, and this can be traced to Pythagoras who proposed total abstinence from beans. He is even said to have been so fanatical about it that he died because he refused to cross a field of beans when being chased by his enemies, who consequently caught and killed him.

The reasons behind Pythagoras's ban on beans have been fiercely argued over the centuries:

Cicero believed that it was because 'beans do not favour mental tranquility'. Plutarch agreed and said that they should be forbidden because they caused wind and inspired bad humours. Aristotle, however, stated that Pythagoras had banned the bean from the diet because it resembled the male testicle. These various views were amalgamated by Richard Taverner, who, in 1539, stated that beans not only caused impure humours but that these humours provoked bodily lust!

Robert Burton, writing on *Anatomy of Melancholy* a few decades later, was not convinced about the lasciviousness, but he too thought that 'the melancholy man should eat no beans'.

The problems for the 'melancholy man' were thought to be even worse when beans were in flower, for it was well known that 'The flowers of the bean, not withstanding that they be of a pleasant and delightsome smell, do hurt a weak brain, and such a one is easily carried away and overcome; and there upon it comes to pass that there are a great number of fooles when beans are in flower.'

This ancient piece of wisdom has been distilled into the saying:

'When beans are in flower, fools are in power.'

BEAN FOLKLORE
AND SUPERSTITIONS

Whether beans and bean plants are regarded as a good or a bad omen depends very much upon where in the world you live.

In Europe beans have largely been held to be unlucky. In Roman times beans were offered as a sacrifice to the god Apollo, but the priests of Jupiter were forbidden to touch them or even mention them. The populace connected beans with death and believed that any black spots on their pale skins was a sure sign of death. Beans are still distributed in Italy to the poor on the anniversary of a death.

In England the flowers of the bean plant were associated with death and accidents were considered much more likely to occur when they were in bloom. To find a white broad bean instead of a green one was an omen of death in the family, and children were discouraged from romping in the bean fields for fear of the bean-goat – a nasty manifestation of the Devil.

During the Jewish Passover certain sects prohibit dried beans, and some extend the ban to green beans as well. However, round beans are used to symbolize the continuity of the life cycle in the Jewish faith and were used whenever the meaning of life had to be explained.

In Japan the bean is venerated, and they are used in a widely observed celebration of ritual purification. Roasted soya beans are used as 'beans of good fortune'. They are scattered in each of the rooms of the house and tossed through an open window to symbolize a rejection of bad luck and a welcoming of good fortune.

TWELFTH NIGHT BEAN FEAST

A very old custom in Western Europe is associated with Twelfth Night. A bean used to be hidden in a special cake and the person who got the slice with the bean was made 'king' of the revels. Originally the king may have reigned for the twelve days of Christmas and his chief function was to perform propitatory rites to ensure good weather. The custom of the bean king probably pre-dates Christianity for Roman children drew lots with beans during the winter Saturnalia to see who would be 'king'.

The origins of the saying 'a bean feast' are attributed by some authorities to the fact that the bean goose was prominent at a blow-out meal. The bean goose had markings on its beak that resembled certain types of beans.

Other experts say that the origins are even older and go back to ancient Greece and the general Pericles. Pericles separated his forces into eight divisions and allowed the divisions to draw lots. The division which drew the white bean was allowed to feast and take their ease, while the others did the fighting.

JACK AND THE BEANSTALK

The pantomime *Jack and the Beanstalk* is the modern version of a fairytale originating way before Christianity. The details vary in different countries, but essentially the story is of a hero who succours his land in times of trouble. The meaning of the magical beanstalk, however, is as lost in the mists of time as the top of the beanstalk is in the clouds!

PRACTICAL BEANS

BEAN BAG TOYS AND CUSHIONS

Small dried beans can be used to stuff soft toys, and indeed this kind of stuffing makes an amusing type of toy which is firm but floppy. Bean bags, made from heavy material, can be made into balls for young children to play with. Larger bean bags can be used as floor cushions and even as beds for cats and dogs.

BAKING PASTRY

Keep a jar of dried beans in the kitchen to use when baking empty pastry cases. Roll out the pastry and line the flan tin or dish. Cover with a piece of foil cut to the size of the base and cover with dried beans. The combination of foil and beans stops the pastry bubbling up and the foil stops the beans sinking into the pastry. Remove the beans and foil towards the end of the cooking time to allow the pastry case to dry out completely.

BEAN JEWELLERY

Drill a hole through the centre of a collection of speckled beans and thread them into a necklace. Alternatively dip the beans in brightly coloured paints or in gold or silver flecked paint.

Beans are also connected with jewellery through the term 'carat'. This term came from carat beans, which all grow to the same size and weight. The natives of the East coast of Africa used them as their standard for weighing gold.